BRITISH BRICKS

David Kitching

AMBERLEY

First published 2016

Amberley Publishing
The Hill, Stroud
Gloucestershire, GL5 4EP

www.amberleybooks.com

British Library Cataloguing in Publication Data.
A catalogue record for this book is available from the British Library.

ISBN 978 1 4456 5912 1 (print)
ISBN 978 1 4456 5913 8 (ebook)

Typesetting and Origination by Amberley Publishing.
Printed in Great Britain.

Contents

Information

Bricks are all around us, and yet we seldom stop to look at them and the structures with which they are built. There is an almost infinite range of bricks and, likewise, the colours in which they can be found. At one time, every small district had its own brickworks to meet local demand, but subsequently larger businesses began to develop, along with better transport links that allowed them to supply regionally. From the single kiln in a field to the massive continuous kilns and chimneys that grew up where the right clays were available, millions of bricks were produced to feed the demands of housing, transport and industry. Specialist requirements for bricks to resist high temperatures were met by using fireclay and silica rock for refractory bricks. Today, there are far fewer producers but their output can be enormous, and modern works continue to supply the demand for the humble brick.

Before the introduction of machinery, bricks were handmade using wooden moulds, usually in small establishments that served local demand, often on a seasonal basis. Development of machinery in the nineteenth century saw significant increases in productivity; there were 109 patents for brickmaking machines and kilns between 1820 and 1850 alone. Inventors such as Henry Clayton and William Percy developed new machines to improve all aspects of brick manufacture and Clayton established a significant manufacturing business. Another company that had a long history of manufacturing brickmaking machinery was C. Whittaker & Co. Ltd of Accrington, who continued in business for over a century.

One approach to mechanisation led to the wire-cut brick, where an extruded column of plastic clay was cut by wires to form the individual bricks. Another process was the use of high-powered presses to force a die onto semi-plastic or dry material in a mould. This latter process would also allow the creation of the recess or 'frog' in the face or both faces of the brick. The die often stamped the name of the brickworks, also known as the 'brickmark', into the frog; it also allowed the patent holder to have their details in the reverse of the brick. Most of the bricks featured in this book are the product of the brick-pressing machines, although there do exist bricks where the name has been hand stamped on to hand-moulded or wire-cut bricks, which do not contain a frog.

Many small, local works are recorded by the name pressed into the bricks that they made, which may now be the only sign that they ever existed. A growing number of enthusiasts have been searching demolition sites and hedge bottoms for examples of bricks from as many brickworks as possible. Sometimes a brick will be found with a name or initials that are completely unknown; and then the long research process begins to try to unravel the history of a long-lost business. Just a small proportion of the brickmarks from British works are shown here, alongside a little of the history of each brickworks featured.

The materials used for making bricks vary greatly according to what is available locally and what particular qualities are required from a brick. Clays of many different types have been traditionally used for brickmaking. Etruria marls found around Stoke-on-Trent make excellent blue engineering bricks, while the shales from Whinney Hill were used to make high-quality red facing bricks in the Accrington area. Fireclay and ganister are used to manufacture refractory bricks for furnace and kiln use, and were commonly extracted by mining rather than from open quarries.

Coal was the usual fuel used to fire bricks, although oil and gas firing became more prevalent in recent years. There was also variety in the form and design of the kilns used to burn the bricks. The simplest clamps sufficed in many places, until demand increased both for quantity and consistent quality of bricks. The round intermittent kiln was very popular, as were a variety of rectangular intermittent types such as the Scotch kiln. More efficient and productive were the various continuous designs, such as the Hoffmann and Staffordshire kilns. In these the structure comprised a series of chambers connected in a continuous loop; the burning would proceed around the loop with chambers in different stages of setting, heating, burning, cooling, and emptying. The kiln never cooled down and the heat of the burning chambers was used to preheat the next ones to be fired. This made for much better energy efficiency than that of the intermittent type.

If you want to see and learn more, there are several museums with collections of bricks to view. These include the Apedale Heritage Centre at Newcastle-under-Lyme, Beamish Museum, The Museum of Lincolnshire Life, Wheal Martyn China Clay Museum, and the Anson Engine Museum at Poynton in Cheshire. In addition, there are a number of websites devoted to bricks, as well as the brick photo albums to view on photographic sharing websites such as Flickr. The whole of Britain is covered on David Sallery's brick site (www.penmorfa.com/bricks) and a comprehensive look at the brickmarks of the Scottish brick industry can be found at www.scottishbrickhistory.co.uk.

Further reading

If you want to find out more about bricks and brickmaking, these are a selection of the books that are available to consult.

Clews, F. H., *Heavy Clay Technology* (Stoke-on-Trent, The British Ceramic Research Association, 1955).

Connolly, Andrew, *Life in the Victorian brickyards of Flintshire and Denbighshire* (Llanrwst, Gwasg Carreg Gwalch, 2003).

Dobson, Edward, *A Rudimentary Treatise on the Manufacture of Brick and Tiles* (London, Crosby, Lockwood and Co, 7th Edition 1882).

Douglas, Graham & Oglethorpe, Miles, *Brick, Tile and Fireclay Industries in Scotland* (Edinburgh, The Royal Commission on the Ancient and Historical Monuments of Scotland, 1993).

Hammond, Martin, *Bricks and Brickmaking* (Princes Risborough, Shire Publications Ltd, 2001).
Hillier, Richard, *Clay That Burns* (London, London Brick Company, 1981).
Woodforde, John, *Bricks to Build a House* (London, Routledge and Kegan Paul, 1976).

Acknowledgements

The preparation of this book has been greatly assisted by a number of people who have been collecting, photographing, and researching bricks and their makers for many years. They have all happily shared their knowledge and images. Particular thanks are due to Ian Castledine, Alan Davies, Martyn Fretwell, Gordon Howle, Alan Hulme, Frank Lawson, Tim Lawton, Ken Perkins, David Sallery, Lawrence Skuse, and Ian Summerfield.

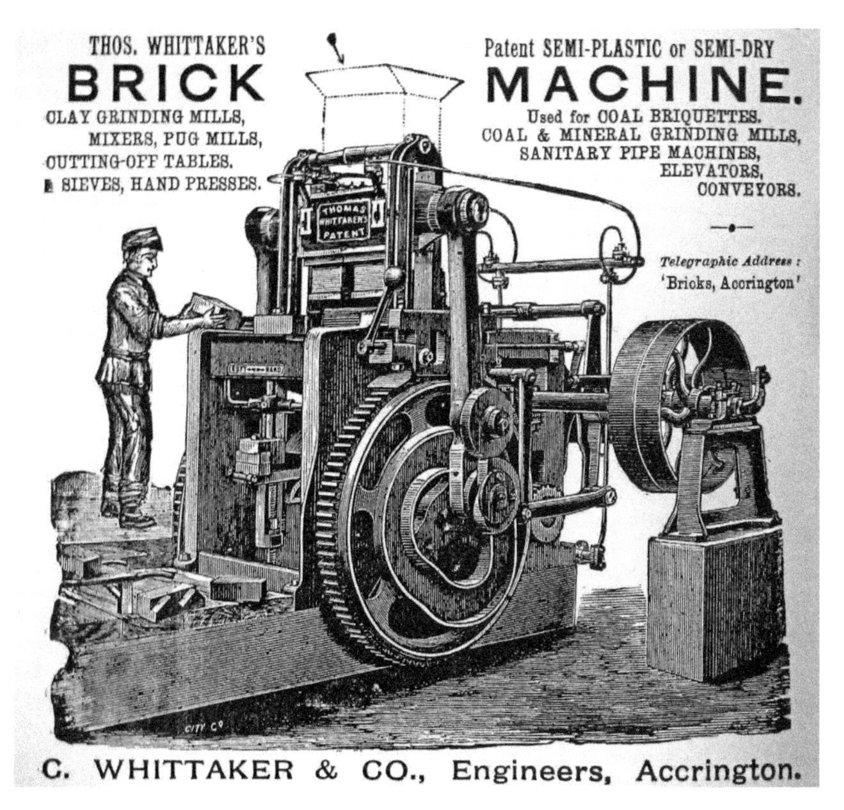

Advertisement for Whittaker's patent brickmaking machine from *The Mechanical Engineering of Collieries*, by C. M. Percy (Wigan, 1890).

Henry Clayton & Co., Atlas Works, London, were the manufacturers of a wide range of patent brick, tile, and pipe machines.

William Carter Stafford Percy, machinist of Manchester, lodged a patent application on 29 April 1847 for 'improvements in machinery for making and dressing bricks and tiles, and in certain sheds and kilns, in which bricks and tiles are dried and burnt', and on 31 July 1858 for 'improvements in arrangements and mechanism or apparatus for the manufacture of bricks, tiles, pipes, and other articles made of plastic earths'. Another, dated 11 December 1861, covered machinery for making bricks, tiles, pipes and other articles formed of plastic materials. It appears that the inventive Mr Percy also spent time in Lancaster Gaol due to bankruptcy in early 1854.

A bronze die, which is typical of the plates that were inserted into brickmaking machines to press the manufacturer's details into the frog of the bricks. This example is from the Vernon Brick Company at Poynton in Cheshire.

A brick press manufactured by James Mitchell & Son of Cambuslang is seen here hard at work at the Furness Brick & Tile Co. Ltd works, producing traditional frogged bricks at a rate of several thousand per hour.

Round intermittent downdraught kiln at Wheatly & Co., Springfields Tileries, Trent Vale, Stoke-on-Trent. Wheatly's was established before 1819, and was famous for its 'Triton'-brand roofing tiles. In 1978 the company was taken over by Daniel Platt Ltd of Brownhills Tileries in Stoke; subsequently in 1982 the works were closed, and demolished soon afterwards.

Hoffmann Kiln at the Colwich Brickworks, Staffordshire. This continuous kiln had six wickets (entrances) on each side and one at each end. Each wicket had an adjacent damper arch, plus four more at the corners, to allow better control of the air passing into the kiln. It is believed to have been constructed by Dean & Hetherington, kiln builders of Accrington, Lancashire.

Belgian Kiln at the Waterside Brickworks, Dunaskin, Ayrshire. This continuous kiln is a variation on the Hoffmann Kiln and was intended to produce less discoloured and overburnt bricks by separating the burning coals from the bricks through using fuel grates across the floor. The dampers and flues are controlled from below, unlike the Hoffmann. It comprises twenty-four chambers and was built in 1935 by William Cleghorn of Newmains.

Ackers Whitley & Co. Ltd were the proprietors of the Bickershaw Collieries west of Leigh, Wigan. When the nearby Abram Colliery closed in 1933, the shafts and brickworks were absorbed by Ackers Whitley. After nationalisation, the brickworks continued under the auspices of the National Coal Board.

Perhaps the most famous trade name in British bricks is the Accrington NORI, which was manufactured at the Altham works of the Accrington Brick & Tile Company, founded in 1887. This high-quality, acid-resistant brick is said to have been used in the foundations of Blackpool Tower and the Empire State Building in New York. The use of the name 'IRON' spelt backwards has had various explanations, including lettering on the works' chimney with the 'I' at the bottom, and accidentally placing the letters backwards in the die for the brick press. Whatever the reason, this name has become synonymous with quality bricks. The business is currently owned by Hanson building products, which has been renamed Forterra.

Adamantine is one of the trade names used by Charles Davison & Co. Ltd, Ewloe Barn Brickworks, Buckley. The 'Made in England' statement is a piece of creative marketing, as this brick was actually made in Wales. The company described this brick thus: 'The adamantine quality can be used for chimney and flue building, and is of particular value in withstanding the action of acid-laden gases. It is also admirable for all kinds of foundation work, manhole linings and culverts. For places where wet conditions exist, adamantine material is eminently suitable.'

Afon Goch & Tatham Tileries, later to become the Tatham Brick & Tile Works, was operated from *c.* 1868 by Henry Richard Bowers & Co., producing bricks, pipes and chimney pots. The works appears to have been taken over by the Ruabon Brick & Terracotta Co. after 1910. By 1937, considerable modernisation had been undertaken and the manager, Mr Packard, had erected the largest round downdraught kiln in the country alongside the first concrete monolithic kiln. Production ceased by the Second World War, when the works were requisitioned as a dispersal warehouse for Bakelite Ltd.

A typical product of the Halifax Glazed Brick Works near Hipperholme, which was once operated by Allen & Son (Halifax) Limited, established by Henry Victor Allen. Many of the fireclay bricks produced here have several faces, with a coloured protective glaze. The works later produced refractory bricks, including the well-respected Selfrac brand, which was perpetuated after a takeover by Scottish manufacturer G. R. Stein.

William Baird & Company, later Bairds & Dalmellington Ltd, operated the Annbank Brickworks, adjacent to their Ayr 1 & 2 (Enterkine 9 & 10) Collieries, which supplied fireclay to the works. The National Coal Board operated the works until 1969, and the Scottish Brick Corporation until 1977.

A group of local businessmen – H. Mellor, G. S. Heath, R. F. Matthews and John Field – formed the Askam Brick Company Ltd in about 1899/1900. The works was constructed during 1900, with plant and machinery installed and railway connections made; brick production commenced, using shale from a nearby quarry, on 1 December 1900. During the period from 1901 to 1902, Furness Brick & Tile Co. Ltd acquired the assets of Askam Brick Company.

The Astley & Tyldesley Collieries Company was formed in 1900 to work the deep pits east of Leigh in Lancashire, formerly owned by Samuel Jackson's Astley & Tyldesley Coal & Salt Company. The brickworks were situated at Nook Colliery, where there was a large continuous kiln.

The Aston Hall Coal & Brick Company works were situated on the slopes below Buckley Mountain. The works were established *c.* 1864 by Messr Kershaw and taken over by the Coal Company in 1868, when there were two kilns in use. This is the face of a Premier brand brick, which were particularly noted for retaining their original colour and for not vegetating, as was the case with the soft, common buff bricks. The colliery closed in 1909 and, as the clay supplies had come from underground, it was just a matter of time before the stockpile of some 10,800 tons was used up. The final dismantlement sale was on 24 July 1913.

Auchlochan collieries at Coalburn, Lanarkshire, were sunk in 1894 and operated by Caprington & Auchlochan Collieries Ltd. The brickworks used the fireclay that was raised alongside the coal and, in 1933, output was 20 million bricks per annum. From the mid-1930s, the collieries and brickworks were absorbed by the ironmasters William Dixon Ltd. It appears that the brickworks had either closed or been sold by 1947, as it was not passed to the NCB.

Sometimes the identity of a brickmaker proves difficult to find even after hours of research. An example of this is this brick that was found during an archaeological excavation at Ashburys close to Manchester City Centre. While the 'B.W' probably stands for '*Brick Works*', the names behind rest of the initials remain a mystery. The brick itself probably dates from the mid-1870s.

H. J. Baldwin Brickworks, Bunny, Nottinghamshire. In 1986 this works, just south of Nottingham, employed three workers in the quarry and thirty-seven in the brickyard. One speciality of this brickworks was the production of covers to protect buried electric cables. After closure the clay pit was used for landfill.

The Balgonie Colliery Company operated the brickworks from the early 1900s alongside the Julian Pit in Thornton, Fife. It appears that the brickworks had been sold by 1947 as it was not passed to the National Coal Board.

James Barker started as the owner of the Ingleton collieries from 1890 and by 1895 erected a brick and tile works adjacent to his Richard and Nellie pits. Transport difficulties and land disputes held back production and the brickworks closed in 1905.

Nathan Barlow was operating a brickworks at Portland Street in Hanley in 1879 and this brick is from when he was operating in partnership with Henry Faija in the late 1860s/early 1870s. Henry Faija was an engineer who came from Newcastle-upon-Tyne in 1868 and became managing partner at the Railway Foundry Company, Stoke-on-Trent. He was engaged in constructing bridges and other works for the North Staffordshire and Market Drayton railways and the demand for bricks to construct these railways may have led Faija into involvement in the brick manufacturing industry with Barlow. He removed to London in 1871 where he set up as an engineer on his own account.

George Basford is first listed in 1864 as a manufacturer of blue and red floor tiles at Ellgreave Street. Basford Brothers brickworks is listed at Dale Hall, Burslem in 1868–70 and in 1873–79 as owned by James Powell Basford. The brickworks was on the south side of Ellgreave Street with a small clay pit by the road and one round kiln.

S. Beardsley & Son operated a brickworks at Ilkeston in the later 1890s. Beardsley lived at 2 St Mary Street, and the brickworks was situated just north of the houses in Mill Street, adjacent to the Erewash Canal.

The Bentley Hall Brick Company Ltd was founded in 1933. The eighty-acre site had previously been extensively mined and much of the material used came from the spoil heaps. In 1938 there was a Hoffman continuous kiln capable of burning over 200,000 bricks a week and three rectangular kilns on the site.

This brick was made at the Longport brickworks owned by the founders of Wilkinson Bros (Burslem) Ltd in the early 1930s. Samuel Wilkinson, the elder of the Wilkinson brothers, had a son named Frank Bernard. Sam decided that the brick's trade name should be 'BERN - ARD', but his brother disagreed when he became aware of this decision. Consequently the dies were destroyed and the brand withdrawn.

Brick and tile making was already established at the Bispham Hall works of William Moorfield in the 1860s. From 1899 the Bispham Hall Colliery Company was operating as a partnership of William Hilton and James Marsden, and then as a limited company from February 1900. In 1925 the Bispham Hall Brick & Terra Cotta Co. Ltd was formed to take over the brickworks. Terracotta production ceased before 1940 and brick manufacture ended in 1973 with earthenware following in 1974.

The Blackstone Edge Brickworks, Whittaker Moor, Littleborough, Lancashire, was also known as Whittaker Pottery. In 1848 Abraham Rogers ran the works as a firebrick and terracotta manufacturer. It was part of James Lees' colliery from at least 1851 and from 1859 - 1863 it was Lees and Fletcher, manufacturers of sanitary tubes, firebricks, chimney tops and terracotta. Closure came in 1873. (Frank Lawson)

Brickmaking at Blaendare, Pontypool, has a long history going back to the eighteenth century. The site was connected with ironworks, coke ovens, and a drift mine as well as bricks. A wide range of firebricks were manufactured by The Blaendare Co. Ltd in 1868 and shortly afterwards by Blaendare Colliery Co. until the early twentieth century. (Lawrence Skuse)

The Bourtreehill Coal Co. Ltd, Dreghorn once operated a mine and brickworks just to the north of Dreghorn village. The brickworks appears to have started around 1875 and ran until 1969.

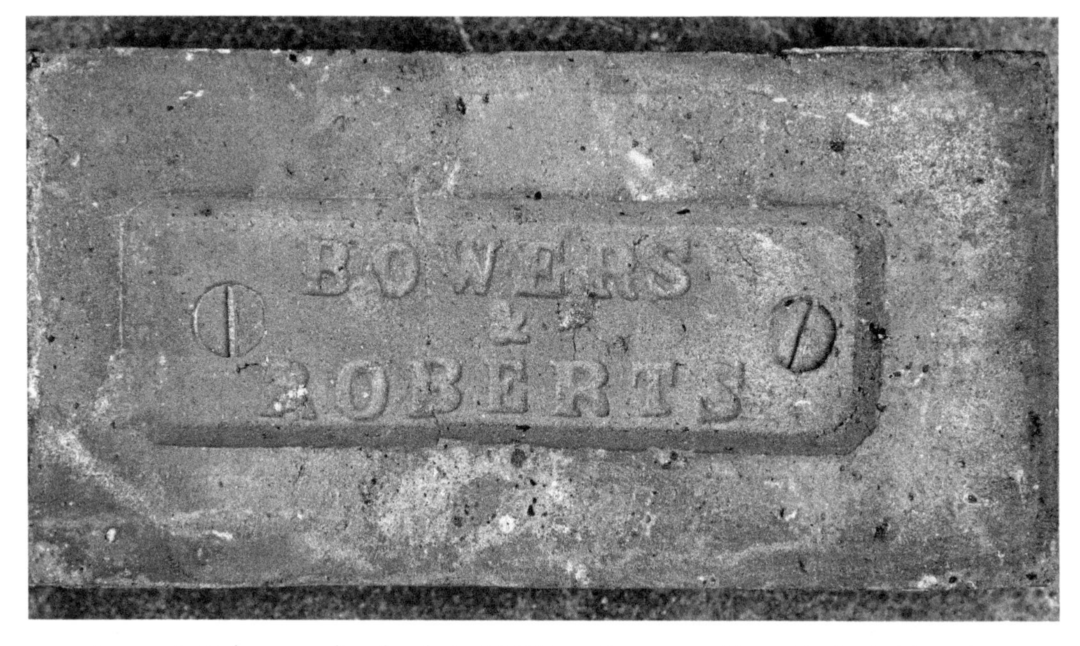

Bowers & Roberts are listed only in 1873/4 at Greengates, Tunstall. The brickworks seems to have been situated on the opposite side of Furlong Road to the Greengates Pottery and in 1869/70 was in the hands of John Meir who was also owner of the pottery. By 1875/6 Anthony Roberts was the proprietor, followed by Charles Salt in 1879 and William Adams & Co. from 1892.

A brick from the H. R. Bowers brickworks at Penbedw, Acrefair. Henry Bowers was operating from this site by 1854 producing glazed stoneware pipes, firebricks and white facing bricks. The business flourished until his death in 1902 but then declined and was liquidated in 1912. The brickworks site was quarried away by Rhos Fireclays Ltd in the 1970s.

In 1852 Mr George Cleunel opened a brick and tile work in Sanquhar, soon building a large and prosperous trade in drainage tiles, but by the 1880s was suffering from a lack of capital and modernisation. In 1889 another tenant, Mr James Brodie, took over and soon improved and extended the works, which by 1891 comprised five Newcastle kilns and a Staffordshire oven. The improved plant included a machine for the production of pressed bricks for outside building. Isherwood Brothers ran the works at a later stage, and then a Mr Scott. Eventually they were taken over by the Sanquhar & Kirkconnel Collieries Company.

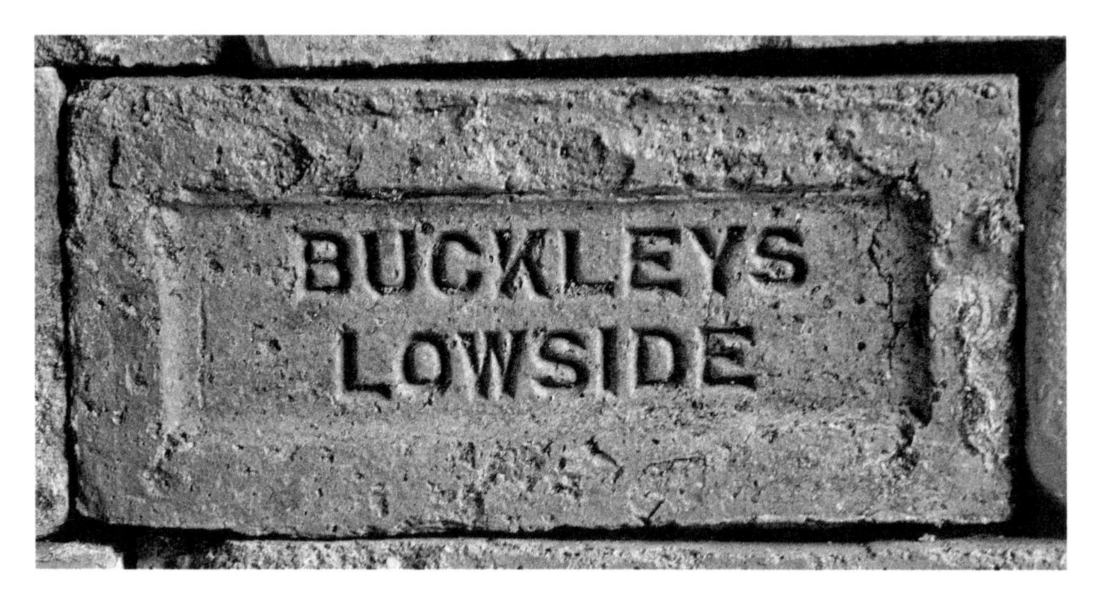

Rothwell and Ephraim Buckley were working the drift mine known as Lowside Colliery on Glodwick Lows, Oldham, from around 1910 until 1938. Immediately to the east was the Lowside Brickworks that appears to have opened after 1900 and was gone by the early 1950s.

John Burgoyne took over the Bryntovey Brick Works by 1881, although George Waters Williams was previously producing bricks in these brick kilns at Llanvihangel-Pontymoile, which were established in 1850. The business also produced drainage pipes. (Lawrence Skuse)

Burthy Brickworks was in the Cornish parish of St Enoder. The works was established by 1887 when T. Nicholls & Co. were proprietors. Machinery at the site was driven by a rotative beam engine. Products included stock bricks, firebricks and tiles for clay dries. The Burthy China Clay Company eventually took over the business and after closure the site was absorbed into the Melbur china clay pit.

Callendar brickworks was to the south of Glen Village near Falkirk. The brick and fireclay works moved to this location in the 1880s as part of the Callendar Coal Co. In 1947, the brickworks were separated off into a private company, the Callendar Brick & Fireclay Co. Ltd. This company was bought by Hepworth Iron Co. in 1967 and the works closed down around 1979. The site had nine kilns of five different designs.

Cannock & Rugeley Collieries had extensive brickworks at Wimblebury Colliery, Littleworth. This was established in the 1870s and was producing more than one million bricks a year in the 1930s, mainly for use in the company's own collieries. Production continued into NCB ownership.

There were two brickworks in Carfin and this is a product of the one operated by Andrew Jeffrey's Carfin Coal & Fireclay Co., Carfin, Motherwell, Lanarkshire, between around 1859 and around 1939. (Frank Lawson)

Carkeet Brickworks near Liskeard used a partly kaolinised granite to make bricks and terracotta. The Terra Cotta & China Clay Co. ran the works from around 1886 for about ten years before the business was abandoned as unprofitable.

The Castle Brick Works was in Birchills, Walsall, and first appears on the 1902 OS map with five rectangular kilns. In 1924 it was operated by J. Griffin, Jones & Company, and in 1940 it was the Castle Brick Co. The works seems to have closed and been swallowed up in extensions to the tube works by the early 1950s.

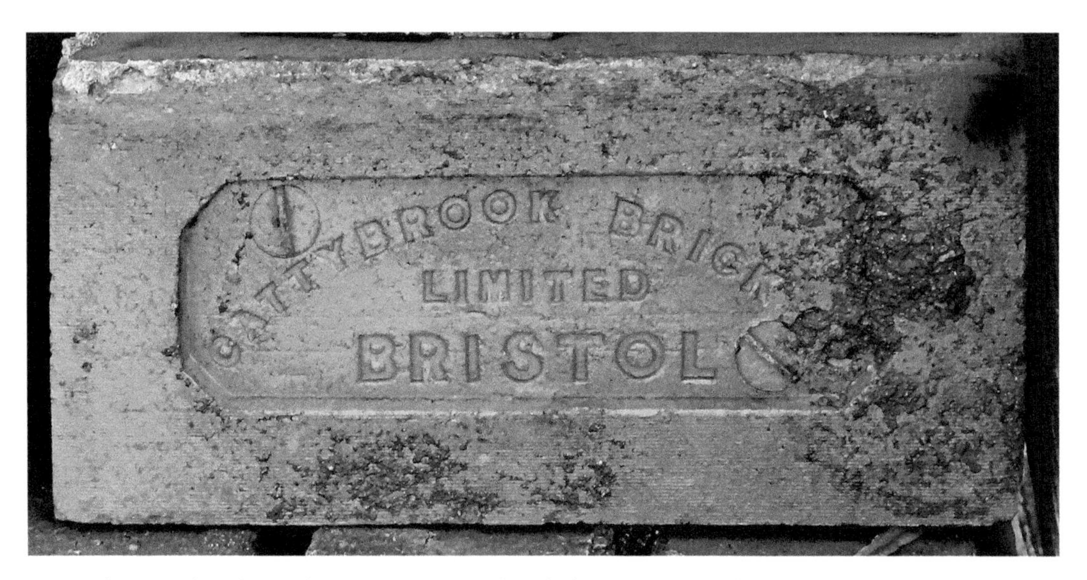

The Cattybrook Brick Company was founded in 1865 in Almondbury, Bristol, and is still operating within the Ibstock Group. It is said that they provided 74,000,000 bricks for the construction of the Severn Railway Tunnel. They also provided bricks for the rapid expansion of the town of Barry in the nineteenth century. (Lawrence Skuse)

The Cheshire Brick Co. works was situated at Middlewood in Cheshire between Hazel Grove and High Lane, Stockport. Formerly trading as the Middlewood Brick Co., the name of the business was changed in August 1932. Two years later a twelve-a-side Hoffman kiln was constructed with finance from the parent company Yorkshire Amalgamated Products. Production ended in the early 1960s and the site was used as a refuse tip.

A small brickworks at Heath House on the edge of Cheddleton, Staffordshire which appears in the 1896 trade directory, but not that for 1904. The proprietor Thomas Clarkson also ran Heath House Farm. The poor quality of the brick and the bits of stone in the structure are typical of the products of such small local brickworks where the clay had very little treatment before being taken to the moulding machine.

The Clay Cross Company was founded in 1837 by George Stephenson the railway pioneer to produce coal, coke, iron ore, iron, and limestone. The company produced their own bricks, with the brickworks originally at the ironworks, but they later moved near to the railway station using modern kilns with a capacity of 5,000,000 bricks per year.

Thomas Cope first appears as a brickmaker in the 1842 trade directory at Steels Nook, Longton, Staffordshire, and the business was listed as being run by his executors by 1864. In 1869 it was known as Holden Bridge Brickyard, Smallthorne, and was last mentioned in 1904.

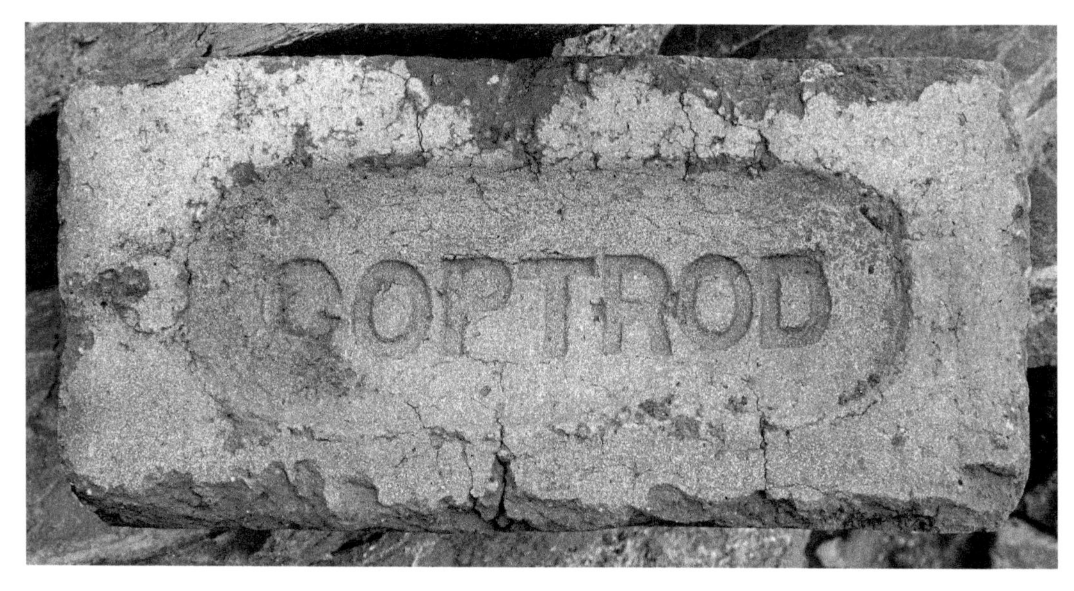

Coptrod brickworks was on Mellor Street in Rochdale, and opened some time in 1908. The business was voluntarily wound up in July 1920 but, by 1924, it was listed as being worked by Summit Brickworks and Sparth, trading jointly as the Federated Brickworks Ltd.

The County Brick & Tile Co. works were established in the 1890s at Stacksteads near Bacup, Lancashire, by Thomas Ratcliffe. It went into liquidation in 1900 and the business was taken over by Henry Heys & Sons but had been closed for some time when the chimney was demolished in 1917.

J. & M. Craig was formed by two brothers, James and Matthew, in 1847, with fireclay and brick works at Dean and Hillhead in Kilmarnock, and at Perceton, near Irvine. They also made high-class sanitary ware. The company was incorporated as J. & M. Craig Ltd in 1896 but went into liquidation in 1923. At both sites the company worked coal and fireclay mines.

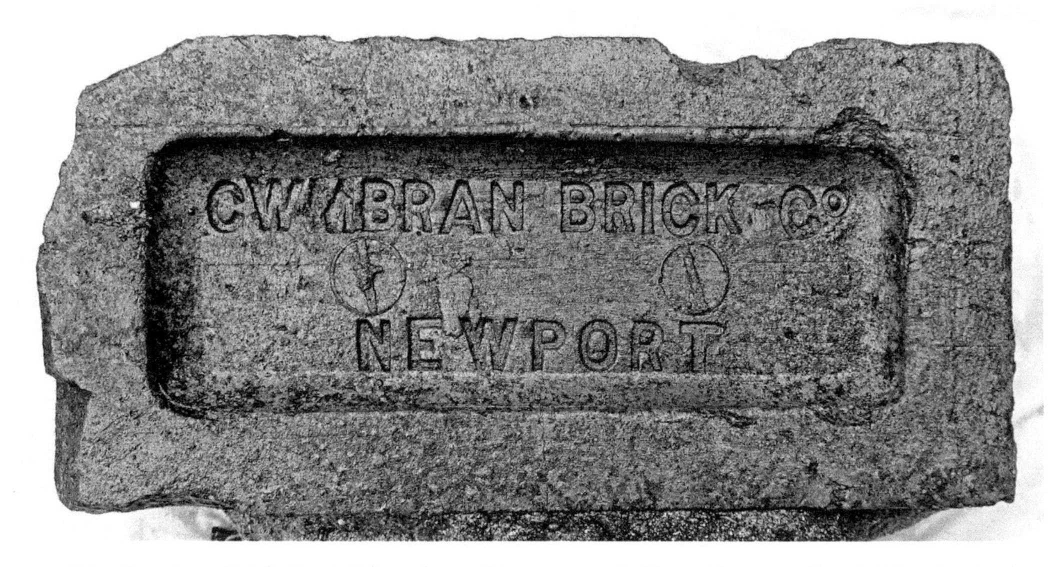

The Cwmbran Brick Co. brickworks at Newport made floor tiles as well as bricks. Production began before 1895. By the 1960s the company was a subsidiary of the Butterley Company and in 1970 was listed as operating at the Llandowlais brickworks. (Lawrence Skuse)

The Cynghordy brickworks were developed by British Anthracite Co. Ltd from the late 1950s to the early 1960s. British Fuel Co. (South Wales) Ltd had acquired the business by 1975 but it is not known if it was still operating at that time. The site was cleared by 1978.

A high-quality blue brick from a Burslem brickworks. Stanley & Rogers were running the works in 1896 and, by 1904, it was Stanley Bros Ltd, Dalehall Brick & Tile Works, Ellgreave St, Burslem. The works were still at work in 1912 but closed by 1924.

Dalquharran Brick & Tile Works were adjacent to the Romilly Colliery site at Wallacetown in the parish of Dailly, in Ayrshire. The works operated from *c.* 1859 to *c.* 1915.

The Darlaston Brick Co. brickworks were situated to the north-west of Darlaston & James Bridge station shortly after 1900. There were six rectangular kilns shown on the 1917 OS map. By 1924, it had become the Darlaston Brick Co. Ltd but, by 1938, the site had been cleared and other industry had developed.

Dewar & Findlay Ltd, Drumpark Brickworks, Bargeddie, Glasgow, which specialised in high quality, pressed facings, either red or blue, and operated from the late 1890s to 1977. Bricks from this works were used in the viaduct at Tarrasfoot on the Langholm branch of the North British Railway in Dumfries & Galloway.

'*Iron*' was a trade name used by James Downing's, Defiance Brickyard, Chesterton, Newcastle-under-Lyme. Downing bought the brickyard from Henshall Moss in 1895 and named it Crown Tileries. His four sons joined the business. James Downing died in 1901 and the works became known as James Downing & Brothers. They went bankrupt in 1907.

The Pocklington Brick & Tile Works in East Yorkshire had a history going back at least as far as the 1850s. Alban Dunwell was brickmaking here in 1905 and, by 1909, the business was listed as James Dunwell (trustees of), Burnby Lane. Production appears to have ceased around the time of the First World War. (Frank Lawson)

The Ebbw Vale Steel, Iron & Coal Co. Limited was a vast business empire that included brickmaking. At Gantre and Brynheulog brickworks, the company used the waste shale from the coal and iron ore mines to make engineering bricks.

Eccleshill Potteries began in 1837 as Manor Potteries, Eccleshill, Bradford, producing salt-glazed stoneware and other domestic items. It was sold to William Woodhead by 1867 and he switched to manufacturing house bricks, firebricks, and sewer pipes until the early twentieth century. (Frank Lawson)

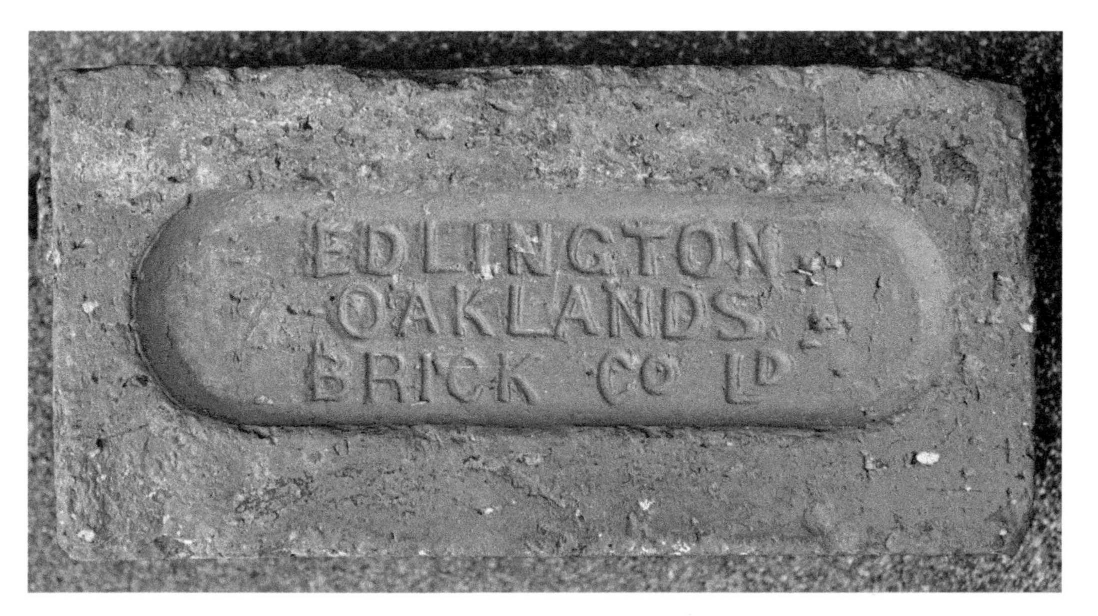

The Edlington Oaklands Brick Co. Ltd erected a brickworks to the west of New Edlington, near Doncaster, by the mid-1920s. Equipment was supplied in the 1930s by engineers Bennett and Sayer of Derby. Brickmaking lasted for a long time at Edlington, and the works were not noted as disused until 1972.

Skiers Spring Brickworks was established in the late 1870s south of Hoyland, adjacent to the branch railway from Elsecar to Lidgett Colliery. In the early twentieth century, Earl Fitzwilliam (EFW) took over the management of the works but these only survived until 1919.

Around 1870 Elijah Forrester Hughes took control of the brickworks at Bleak Hill, Cobridge, Stoke-on-Trent, which had been started by his father John in the 1840s. In 1881 the census shows him as a firebrick manufacturer employing eleven men. He moved his home from Burslem to Little Haywood by 1891 and the works seem to have closed around 1896.

The Eltringham Sanitary Pipe & Brick Co., Eltringham, Prudhoe, Northumberland. Previously, to 1891, this works was run by Harriman & Co. of Blaydon, and the new company was established by their former manager. A narrow gauge railway with a rope haulage system, moving six tubs at a time, brought fireclay from West Mickley Colliery. Production of salt-glazed bricks and pipes was a speciality at Eltringham and production only ended in 1975. (Frank Lawson)

The Etna Works at Bathville, near Armadale, began operation in the 1860s and, among other fireclay products, produced a common building brick. The works were modernised in the 1950s, and new kilns built in the 1960s. Production of red common bricks continued into the 1990s.

Glebe Colliery was established by the mid-nineteenth century and operated for around 100 years. From the 1860s it was operated by Challinor & Co., and then, by 1900, J. Heath & Co. In 1919 it was in the hands of Fenton Collieries Ltd, and continued to work into the nationalised era. It finally closed under the National Coal Board in October 1964. The associated brickworks were situated a short distance to the south of the pit and had three round kilns in 1878. The brickworks were still in business in the 1930s, but had been levelled by 1953.

Joseph Firbank was a contractor who was responsible for the building of railways in many parts of Britain in the nineteenth century. This brick came from a bridge on the Lewes & East Grinstead Railway, better known today as the Bluebell Railway. Firbank was the contractor for this line, which opened in 1882, and it is likely that he opened a works to produce bricks specifically for this job.

Found outside a house in Wellington Street, Whittington near Chesterfield, this shows a misspelling of the name in the frog. This is a product of Jno. Froggatt of North Wingfield, Derbyshire, listed as a brickmaker in the 1857 White's Directory of Derbyshire.

The Furness Brick & Tile Co. Ltd has been established for over 145 years at Askam in Furness and remains a family-owned business. The works continue to manufacture a vast range of standard and bespoke bricks to order.

Gadbury brickworks, on the south side of Wigan Road from Gibfield Colliery, was named after the old Gadbury Fold Farm nearby. It was established by Fletcher Burrows & Co., owners of Atherton Collieries around 1914. The works closed around 1964, shortly after the colliery finished working.

In 1870, William Hammond went into partnership with his brother-in-law Robert Gardiner to work fireclay in Pott Shrigley, Cheshire, and they established their brickworks on the north side of the Bakestonedale road, opposite that of George Lambert. 1875 saw Mr Lawrence Gardiner succeed his brother in the business. He died in 1886 and Hammond took over the works entirely; so this part-glazed brick must date from 1870–1886.

These brickworks were situated at Auchinlee Farm in Bellside, just north of Coltness in Lanarkshire. It was well established by the late 1890s but had disappeared by 1920.

In 1906, Gregory & Co. were brickmakers at Pentwyn, Abersychan, which opened before 1902. By 1926, the Abersychan Brick Co. is listed as producing bricks at Pentwyn. Jabez Gregory also owned the Robin Hood pub at Pontnewynydd and the Pentwyn coal mine. (Lawrence Skuse)

This is one of the more decorative designs that have appeared on bricks. Francis Harry Gordon opened his brickworks in the 1870s, on a site with a 30-foot-thick bed of clay, adjacent to the Daw End Canal just south-west of Clayhanger Bridge at Brownhills, Staffordshire. The buildings included three drying sheds, the largest measuring 150 feet by 30 feet, having a cast-iron-plated floor. Nearby, on the other side of the canal, was the Walsall Wood Colliery brickworks, and price cutting by these works led to the demise of the Gordon business in 1896.

William Gradwell was a major building contractor in Barrow-in-Furness. In 1855 he moved his business to Hindpool, Barrow, and set up a brickworks in Dalton Road that was capable of turning out 75,000 bricks a week.

It is believed that the Grampound Road Brick & Tile Company ran these brickworks, situated between Truro and St Austell. Production commenced *c.* 1880 and, in 1889, the company was advertising, 'Red facing bricks and Plain Pressed Vitrified Blue bricks for foundations, sewage works &c'. By 1907 the works are shown as disused.

The Grane brickworks opened in September 1895. A newspaper report of the opening described the plant:

> The present works consist of a boiler house, engine house, machine house, and grinding house. They are erected in red brick, and are easily accessible, facing the turnpike. The boiler is 37 ft. by 7 ft. No.1 Grinding mill is 9 ft. in diameter. The engines, which can be worked up to 90 horse power, are by Bellhouse of Manchester. The machinery is of the latest patent and is applied by Messrs. C. Whittaker & Co. of Accrington. It is expected as the works now stand, a turn out of 500 bricks a hour can be effected. There is now only one machine, but another four are expected.

It is perhaps unsurprising that the Great Western Railway produced its own bricks, as it would have always had building works underway somewhere on its system. In 1886 there was a brickworks with three rectangular kilns within the railway works at Swindon.

Pilkington's owned a number of collieries in the St Helens area and also the ancilllary brickworks. Greengate Colliery was at one time owned by the Greengate Brick & Tile Co. The colliery closed in 1915 but the brickworks appear to have continued after this. It was situated close to the railway on land at the rear of the current Pilkington's Greengate factory.

Enoch Hampton was making bricks at Eastwood Vale in 1851, and E. Hampton & Sons are listed in 1896 as operating Eastwood & Mousecroft Fire-brick and Marl Works, Hanley. The company was still operating in 1924 but had disappeared by 1940.

Cyrus Hanson of Cwmbran started a brickworks in 1842 for the production of firebricks. He also owned the Henllys Colliery, which supplemented fireclay found on site. The colliery and brickworks were sold in 1874 to wire manufacturer J. C. Hill, who continued to make bricks with the Hanson brickmark at what was then known as Henllis Firebrick & Retort Works, Llanvihangel-Llantarnam. Later ownership passed to Guest Keen & Nettlefold and the brickworks continued until the 1980s. (Lawrence Skuse)

Hawkins Colliery was also known as Cannock Old Coppice Colliery and commenced working *c.* 1840. Joseph Hawkins took on the lease in 1869 and the colliery remained under the control of the Hawkins family until nationalisation in 1947. The company also operated a brick and tile works with a clay pit at Longhouse, adjacent to the Wyrley & Essington Canal. By the 1920s Henry Hawkins was advertising a wide range of products from the Longhouse Brick & Tile Works, Cannock. About 7 million bricks were made annually, using clay mixed with mining spoil. The works continued in Hawkins's ownership after the colliery was nationalised, and remain in production today under the Eternit name.

William Hewitt, Bradwell Hall Brick & Tile Works, Chesterton, Newcastle-under-Lyme. Hewitt appears in the trade directories between 1904 and 1912 at Bradwell Hall, where he advertised as manufacturing bricks and tiles, along with ornamental roofing, ridge and garden tiles.

Hexter, Humpherson & Co. Ltd, The Potteries & Brick & Tile Works, Newton Abbot, South Devon. The business appears to have been established in 1890. Their 1902 catalogue describes the firm as:

> Manufacturers of plain & ornamental terracotta, vitreous engineering bricks, vitreous buff pavements, real stoneware pipes and traps, tested pipes, patent hassall jointed pipes also stanford jointed pipes, terracotta chimney tops, garden edging, real stoneware rustic & grape vases, firebricks, of exceptional quality.

Henry Heys was a quarry owner who came to Stacksteads, near Bacup, Lancashire, from Helmshore in about 1848. By the late nineteenth century he had quarries at Brandwood, Facit, and Hambeldon. His sons took over the business on Henry's death and, as Henry Heys & Sons, took over the brickworks at Rakehead from the County Brick & Tile Co. On the death of the eldest son in 1902, the firm was constituted as a limited company, Henry Heys & Co. Ltd. The brickworks did not operate for long and, in 1917, when the chimney was demolished, had already been disused for some time.

This distinctive brickmark has a certain style. Joseph Higson & Company's brickworks on Bella Street, Daubhill, Bolton, were established in 1873. The brickworks had a working life of eighty-five years, being demolished in 1958.

The Ravenhead Sanitary Pipe & Brick Co. was located alongside Burtonhead Road, in between Ravenhead Colliery's Nos 7 and 8 pits and Nos 9, 10 and 11 pits. It had been established in around 1850 by W. Edwards, David Horn and John William Kelly, trading as Lavender & Co., with premises at Ravenhead Pottery and in Liverpool. In May 1857 Edwards left the partnership and the firm became Horn & Kelly, and later the Ravenhead Sanitary Brick Company. In May 1874 their partnership was dissolved, with David Horn appearing to have taken over the business; in 1875 it was registered as the Ravenhead Sanitary Pipe & Brick Co.

J. & R. Howie Ltd, Hurlford Fireclay Works, Kilmarnock. The works were just to the east of Kilmarnock and was supplied with fireclay by a number of mines in the area. It operated from 1858 until 1971, producing pottery, bricks, chimneys, garden ornaments and enamelled sanitary ware. The buildings were demolished by 1979.

J. & A. Jackson Ltd was formed through the amalgamation of the businesses of two established brickmaking familes in the Manchester area. The Harrisons had four works around Stockport, while the Jacksons operated from three works in Chorlton, Longsight, and Levenshulme. Over the next fifty years, the company expanded with the acquisition of works across the north-west and into the west Midlands. In 1973 Christian Salvesen acquired the business and, in 1995, a management buyout of the brick side of the business saw the formation of Chelwood Brick, which was later acquired by Wienerberger. The business continues to operate from the works at Denton.

Buckley Junction brickworks were established by John Jones and Henry Lamb in 1911, when they took on a lease of the site and set about working the clay. In 1919 Frederick Phelp Jones acquired the business and renamed it the Buckley Junction Metallic Brick Co. Ltd. The company used the trade marks 'JACOBEAN' and 'CITY' for some of their facing brick products. In 1956, after the company became bankrupt, the works were acquired by the Castle Firebrick Co., but this only lasted for three years before complete closure.

Joseph Joberns & Co. is listed as operating the Aldridge blue tile works from at least the early 1880s and, in 1940, had become Joberns Ltd. The works were also known as Coppice Lane Tileries, Walsall Wood. The two trumpet-like decorations stamped either side of the company name are an unusual addition.

This is quite an early lettered brick, having been made at William Jones's Springfield Tileries, Trent Vale, Stoke-on-Trent. Jones is only listed in the trade directories between 1865 and 1870. (Frank Lawson)

The Pelaw Terra Cotta Works were established in 1895 by Jones and Maxwell. By 1911 Jones Brothers owned the yard and manufactured terracotta, engineering, and facing bricks on a large scale. It closed in 1968. (Frank Lawson)

Fauldhead Colliery at Kirkconnel was easily the largest pit in Dumfriesshire. Brickworks were opened here in 1912 by Sanquhar & Kirkconnel Collieries Limited. The National Coal Board took over in 1947, and the works remained in production until a few years after the closure of the colliery in 1969.

The Kirkhouse Brick & Tile Works Ltd, Kirkhouse, Brampton, Cumbria. These brickworks were erected in 1926/7 on land incorporating the old locomotive shed on Lord Carlisle's Railway. Shale was brought from Foresthead limestone quarry, where it formed part of the overburden. There were three kilns and two brickmaking machines by Bradley and Craven. Production increased after 1945, but the business went into liquidation in September 1972. (Frank Lawson)

Joe Kitson & Sons Ltd, Denby Dale, West Yorks. These brickworks, which also produced glazed pipes, appear in a trade directory for 1881. The brickworks suffered a disastrous fire on 23 August 1915 but the business survived and, by 1929, there were eleven round kilns on site. The company was liquidated in 1978. (Frank Lawson)

The London & North Western Railway made almost everything it needed at its Crewe works, including bricks. The 1876 1:2,500 OS map shows a large round kiln near to West Street in Crewe, along with rectangular kilns within the works buildings, adjacent to the Chester line. By 1898 the rectangular kilns had gone and the round kiln had a standard-gauge siding entirely circling the structure. There were also four more smaller round kilns nearby.

In around 1820, George Lambert and Abraham Bury started a small brickworks near Brink Farm, Rainow, which was near Macclesfield in Cheshire, with one kiln to fire the bricks. When they needed to expand, they moved to a site in nearby Bakestonedale. By 1848 George Lambert was operating a coal and fireclay mine with associated Pott Brickworks on the south side of the road at Bakestonedale, Pott Shrigley. Lambert was still working the Pott Shrigley coal mine in 1884, but had relinquished the firebrick works to James Hall before 1878. This firebrick must date from between 1848 and 1878.

The Lanemark Coal Company opened its brickworks in the early 1890s adjacent to its Boigside No. 2 Pit, just north-west of New Cumnock in Ayrshire. The company collapsed in 1908 and the works were closed, although the collieries were taken over by New Cumnock Collieries Ltd in 1909.

William Lea, of Apedale Road, Chesterton, Newcastle-under-Lyme, advertised as a brickmaker, manufacturer of ridge, roof and floor tiles etc., and also as a builder. This brick was used in building the railway station at Silverdale in 1865. The works seem to have closed in the early 1870s.

Littlemill Colliery and Brickworks at Rankinston, Ayrshire, were owned by the Colyton Coal Co. until 1937, and then by Bairds & Dalmellington Ltd until 1947. The works passed to the National Coal Board in 1947 and was closed by 1963.

This works at Little Mill, Pontypool, was known as the Bryntovey Brick Works when established in 1850. From 1881 to at least 1901, John Burgoyne was proprietor, followed by David John Lougher. By 1922, the Little Mill Brick Co. was operating here with a large continuous kiln, and production continued until the 1980s. (Lawrence Skuse)

Kersland Brickworks, Dalry, Ayrshire, was built on the site of the former Kersland Ironstone Pit, and operated from around 1900 to around 1939. In 1903, this and the nearby Carsehead works were run by the Ayrshire Brick & Fire Clay Co. Limited.

William Longmore's, Hopyard Brick Works, Bentley, Walsall, was situated alongside the Bentley Canal. It opened around 1900 with four rectangular kilns producing pipes as well as bricks. William died in 1913 and the works passed to a partnership of his three sons, who worked it until it closed around 1946.

Adam Mason owned Montcliffe Colliery at Horwich, near Bolton, in the nineteenth century and used the fireclay that was found with the coal seams to produce firebricks, trading as Adam Mason & Sons. In 1886 he supplied bricks to build the police station in Horwich. Fireclay continued to be produced by Mason's until the colliery closed in 1966.

John McDonald, Son, & Company, Composition Brick and Fire Clay Manufacturers, Meadow Brick Works, Braidwood, Carluke, South Lanarkshire operated from 1889 until 1920. The business partnership between Charles McDonald and William Barr was dissolved on 30 April 1910. It was also known as the Nellfield Brick & Tile Works.

Micklam brickworks at Lowca also produced large quantities of refractory bricks for the steel industry. In the 1930s it was owned by the United Steel Company Ltd. Clay was worked from a number of drifts, including the nearby Micklam fireclay mine. Much of the machinery for the works was supplied by brick machinery specialists C. Whittaker & Co. Ltd of Accrington.

Milton Brick & Tile Works, Carluke, Lanarkshire, operated from around 1897 to 1978. Scottish Clay Hollow-Ware Limited were incorporated on 15 July 1937 with their registered office at the works. Final closure came on 14 August 1978, with the voluntary liquidation of T. B. Gibson & Co. Ltd, trading as Milton Tile Works, Carluke.

Monk and Newell were Liverpool-based contractors who, in 1883, established their brickworks in Ruabon. The 1899 OS map shows twenty kilns and extensive buildings, with a chain-hauled tramroad leading from the adjacent clay pit. Thomas Monk died in 1889 and John Newell continued to operate the works. The business had closed by 1917, when the works was requisitioned by the Ministry of Munitions; the lease was sold to Thomas Dugdale Stubbs, who sold it on to his own company, The North Wales Brick & Tile Co., which operated until 1929, manufacturing under the trade name 'RUBRIC'.

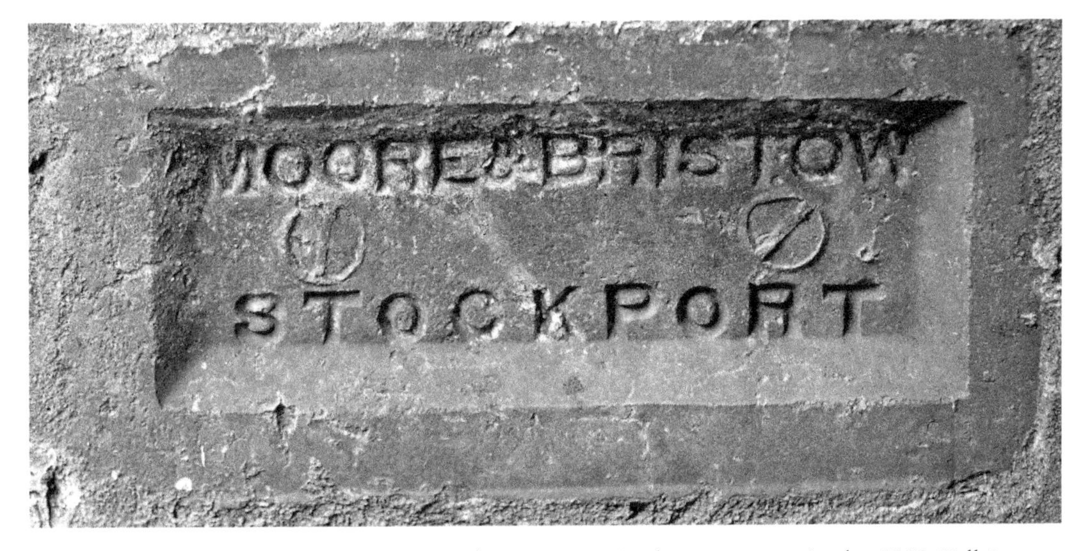

The brickworks of Moore & Bristow, Dialstone Lane, Stockport, appear in the 1902 Kelly's Directory, but not those of 1896 or 1906, making this likely to have been a short-lived venture. The 1907 OS map shows a brickworks with two round kilns at Stepping Hill on land just to the east of Dialstone Lane.

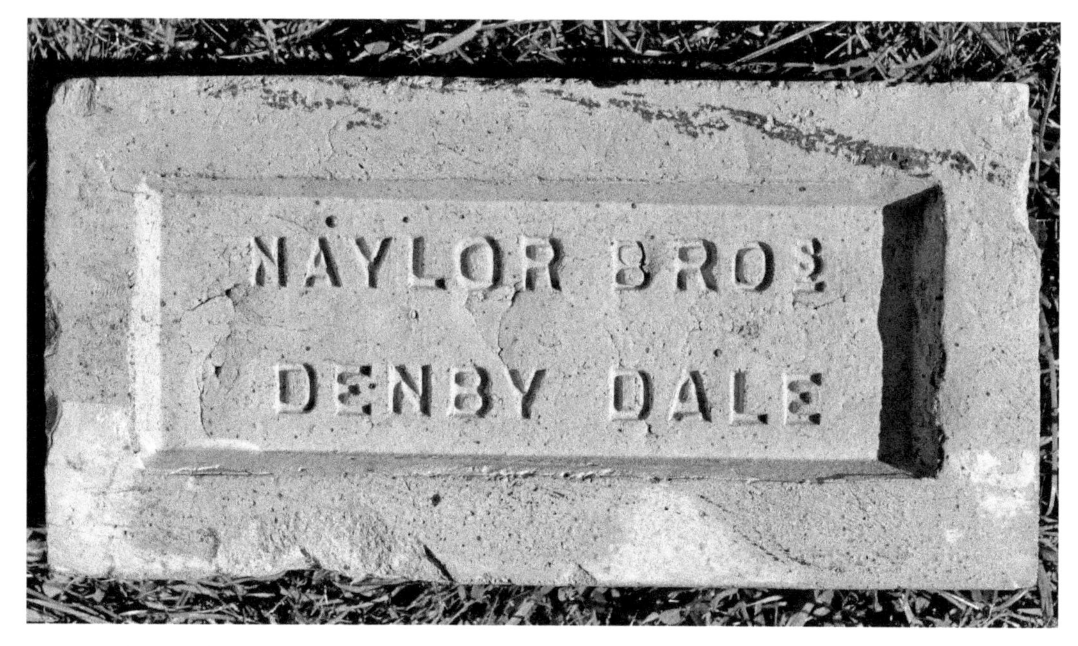

Naylor Brothers at Denby Dale were firebrick manufacturers, although their principal product was at one time salt-glazed earthenware pipes. The works opened after 1891 and, by 1903, there were fifteen round kilns there. Clay pipes were added to the products using material from nearby clay pits. Production continued until 1993, when it was moved to Cawthorne. (Frank Lawson)

New Haden Colliery was acquired by John Slater in 1917 and became part of his Berry Hill group in 1922. A brickworks were opened at New Haden to use clay worked from measures adjacent to the Little Dilhorne seam at the colliery. It was still operating in 1947, although the colliery had closed in 1943. By this time the works were using pit shale from the tips as a raw material.

A brick made at the Norbury Colliery brickworks operated by Messrs Clayton & Brooke in the nineteenth century. The works was located behind the Robin Hood pub on the north side of the main road from the colliery, which was situated between Hazel Grove and High Lane near Stockport. With the closure of the colliery in 1892, it is thought that Moore & Bristow from Stockport may have operated the works for a short period.

The Norbury Moor brickyard was situated on Jackson's Lane in Hazel Grove, Stockport, and appears to have started production around 1890. It is shown on the 1907 surveyed 25-inch OS map, but probably didn't survive the First World War. The map shows a large preparation building and four round kilns.

The North & South Buckley Colliery, Brick & Tile Co. Ltd was incorporated in 1881 and absorbed the South Buckley Coal & Firebrick Co. and the North Buckley Colliery & Firebrick Co. The South Buckley works flourished and expanded to the extent that they outgrew the collieries; they were renamed the South Buckley Rock Brick Co. in the early 1890s.

The North Cornwall Brick & Tile Co. Ltd, Tolcarne, Newquay. The Tolcarne brickworks were probably opened in 1886 and were equipped to produce 10,000 bricks per day using a semi-dry brick machine, manufactured by Henry Tuke & Sons of Leeds. The raw material was silt from the bed of the River Gannel. Production ended in 1896, when the works were unsuccessfully advertised for sale.

Obsidianite was one of the trade names used by Charles Davison & Co., Ewloe Barn Brickworks, Buckley. One of their catalogues states, 'There appears to be in the USA no product equivalent to Obsidianite manufactured by Charles Davison Co of Buckley, England'. This example was seen at the Shotton Steelworks.

Oldfield Colliery in Fenton and its associated brickworks was owned by Balfour & Co. in the 1880s.The colliery passed on to another company, the Lane End Works Ltd, by 1889. Then in September 1896 the colliery, now in the hands of the Oldfield Colliery Company, was closed, leaving the brickworks to operate in their own right. This yard had a sustained life through until at least 1959, when it was manufacturing refractory bricks under the ownership of D. Duddell Ltd.

In 1867, Henry Parfitt leased the Green Meadow brickworks off Mount Pleasant Road, Cwmbran, from John Lawrence. The initial lease was for seven years at £114 per annum, plus 1*s* 6*d* per ton of clay taken from premises. Parfitt continued to produce both red bricks and firebricks here until 1896, by which time he was in rent arrears. The works passed to Guest Keen & Nettlefold some years after Parfitt died in the late 1890s. (Lawrence Skuse)

Mud dug at low tide from the harbour area was brought to the quay by barge for use at the Par Harbour brickworks. The first attempts from 1836 failed to make successful bricks from the mud, and it was not until the proprietors, Messrs Treffry, Clunes & Co., employed one John Alderman, locally known as 'Bricky Jack' that the business began to flourish. In 1928 the works were offered for sale with a capacity of 8,000 bricks per day. By 1935 the site had been cleared.

Pear Tree Glazed Brick & Marl Co., High St, Hanley, Stoke-on-Trent. A short-lived operation that only appears in the 1907 trade directory. The name derives from the Peartree Pit of the Shelton Collieries, which had previously occupied the site. High Street in Hanley is now named Old Town Road.

The Potteries Brick Co. Ltd acted as a selling agent for at least eleven brick companies in the area. Each brickworks used a different letter on their bricks, but all stamped them with the PB Co. label. The works involved in 1943 were: Beans Brick & Marl Co., Tunstall; Berry Hill Brickworks Ltd, Fenton & Cheadle; Birchenwood Brick & Tile Co. Ltd, Kidsgrove; Cobridge Brick & Marl Co. Ltd, Cobridge; D. Duddell, Fenton; J. Hewitt & Son (Fenton) Ltd, Fenton Low; J. Leigh & Sons, Burslem; Wm. Palmer, Exors of, Cobridge; Sneyd Collieries Ltd, Burslem; The Stafford Coal & Iron Co. Ltd, Great Fenton; Thos. W. Ward Ltd, Apedale, Chesterton.

Located at the eastern end of Porth beach near Newquay, this brickworks were owned by Martyn & Bennett from 1880 until John Argall took over *c.* 1909. Materials for the bricks were obtained from a decomposed vein of quartz-porphyry.

Poynton, Cheshire, once had an extensive coal mining industry, owned by Lord Vernon, which lasted until 1935. Ancillary to the colliery was a brickworks located next to the railway. This works comprised a steam engine with a 16-inch cylinder, boiler, crushing and grinding machinery, pug mill and brickmaking machine, supplied by Easton & Tattersall of Leeds. Capacity was 2 million bricks per annum, which were burned in a patent Pollock & Mitchell brick kiln with eight chambers. The brickworks had closed by 1920, when it was offered for sale along with most of the Vernon estate.

George Procter was a farmer at Copshurst Farm at Lightwood, between Normacot and Meir Heath, Stoke-on-Trent. He is listed as a brick manufacturer at Coshurst from 1864 until 1868 and from 1869, trading as Procter & Benbow with his son-in-law Charles Benbow. In 1871 they were employing seven men and six boys at the brickworks. After Procter's death, Charles Benbow continued trading on his own from 1873 until 1904.

Red Hill Bank Brickworks were established in the 1890s, alongside another brickworks, the Rocester Works adjacent to Rocester station, Staffordshire. By 1922, the two establishments appear to have amalgamated into one works, which was still in business in the mid-1950s. The site is now occupied by the JCB factory.

A glazed product of the South Buckley Rock Brick Co., which operated under this name from the early 1890s until the closure of the works in 1914. Fireclay was supplied, along with coal for the kilns, from the adjacent colliery. When the colliery closed in 1909, the clay and coal were brought by a tramroad from the associated North Buckley site, which had been renamed as West Buckley Colliery by this time.

The Wilderness Brickworks at Gresford were established by Edward Stanley Clark in 1885 and, soon after, he was joined in partnership by Russell and James Rea and Charles William Massey, all from Liverpool. In 1888 they formed Clark & Rea Ltd, and became known for producing quality red and buff bricks. In 1903 Edward Stuart Clark became the sole proprietor of the company, which was then operating eight round kilns for red bricks and seven square kilns for blues. It appears that the works were never profitable and that the Clark family lost money for a number of years. It closed by 1924 and was dismantled in 1926.

William Rowland was operating a brickyard at Birch Lane, Dukinfield, in 1857. The letters on this brick are those of John Rowland, Dukinfield, who was William's father, and who died in 1856.

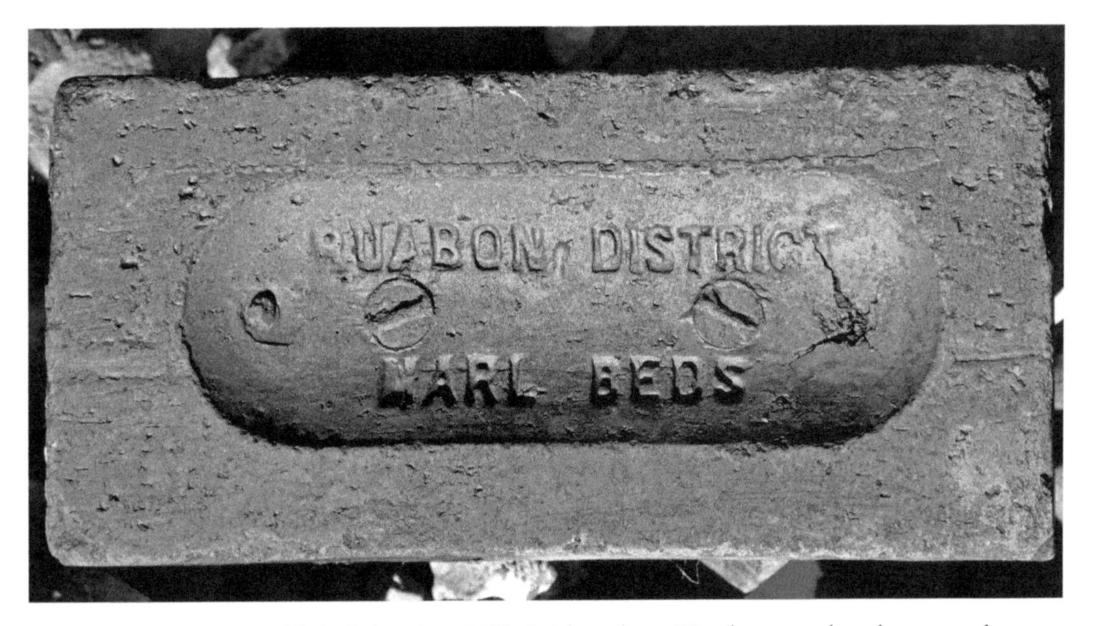

In 1884, E. M. Jones established the Kings Mills Brickworks at Wrexham to take advantage of a bed of superior Ruabon marl. Messrs Phillips & Whitehouse took over in 1895 and they formed The Wrexham Brick & Tile Co. in an attempt to raise capital in 1903. The business failed in 1907 and so The Wrexham Brick & Tile Co. Ltd was formed in 1909 by a new syndicate of owners. Production continued until 1949, although after 1930 this was in the ownership of the Standard brickworks at Buckley.

The Buccleuch Terracotta Brick Works at Sanquhar were initially operated from at least 1867; at one time this was by James Irving McConnell, followed in 1903 by the Sanquhar & Kirkconnel Coal Company. From 1925 it was part of William Baird & Co. Ltd, then Bairds & Dalmellington Ltd from 1931 to 1947, before being closed by the National Coal Board in 1958.

These Cornish brickworks were operated under several similar company names, such as the St Day Firebrick & China Clay Co. Ltd. Originally opened in around 1860 by a Mr Hawke, it then passed to a Captain Nettle of Truro. There was a hexagonal Hoffman kiln and two round kilns, along with extensive clay pits. Closure took place in 1912, although clay extraction continued for a few more years.

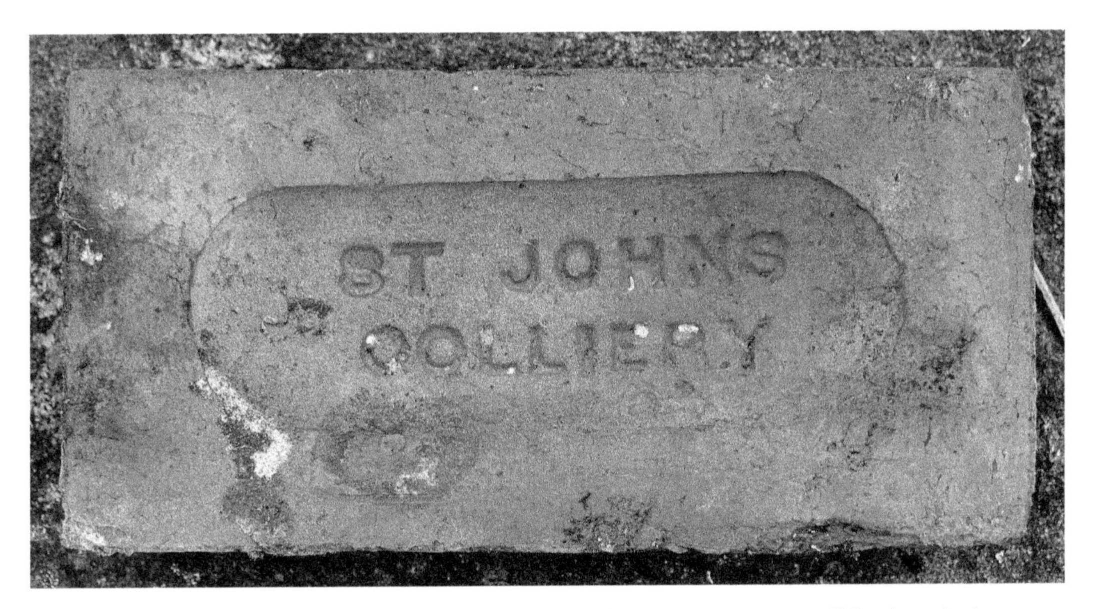

St John's Colliery was at Normanton, near Wakefield, and was operating a small brickworks just south of Newland Lane in 1892. By 1907 the works had been moved to a site on the north side of the road. It survived into the 1970s, but had been demolished by 1985.

In 1881, these brickworks were listed in a trade directory as run by G. Howard, St Julian's, Christchurch, Newport. By 1895 the St Julian's Brick & Tile Co. Ltd, Newport, was listed as 'manufacturers of wire-cut, pressed and ornamental bricks &c'. The works were still in business in 1926. (Lawrence Skuse)

The Seiont brickworks were established around 1850 and were being operated by William Hayward in 1874. Bought by John Summers & Sons Ltd of Shotton, it traded as the Castle Brick Co. Ltd until purchased by Hanson in 1972. This was the last brickworks in Caernarvon and finally closed in 2008.

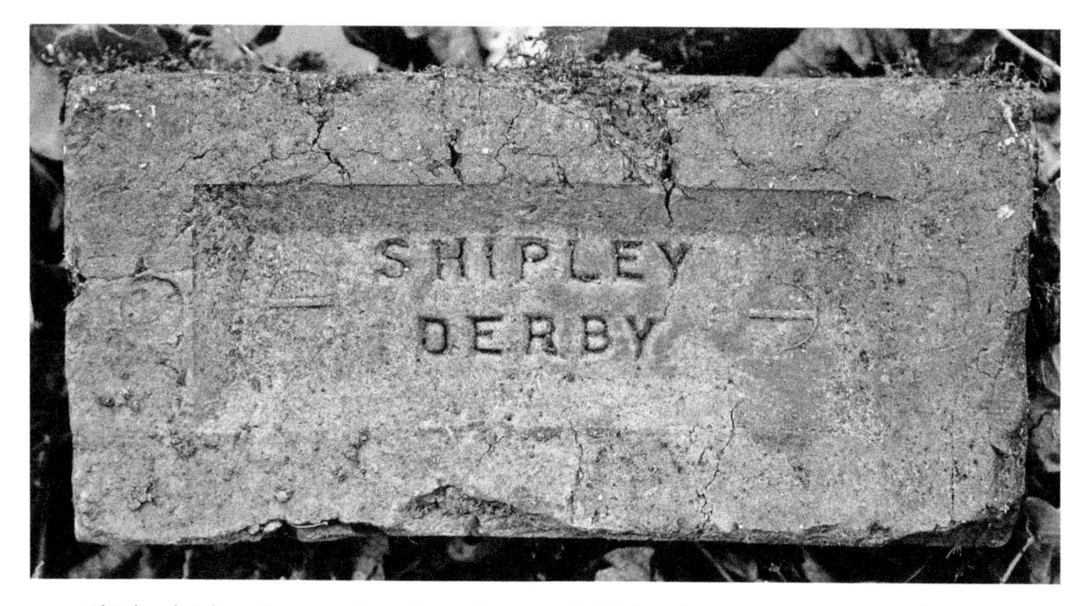

Shipley brickworks was situated north-west of Shipley Common, adjacent to the railway at Shipley wharf at the terminus of the Nutbrook Canal. It was operating in the 1880s and had grown by 1900, but had gone by 1914.

Sneyd Colliery & Brickworks Company at Burslem, Stoke-on-Trent, was operated by Messrs C. and J. May from 1844. In 1875, it was purchased by Messrs William Heath, Arthur Dean and W. A. M. Telwright, and was converted into a limited liability company in 1881. Glazed bricks were a speciality of this works, which traded into the 1960s.

Made at the Graigddu works at Cwmnantddu. In 1881, Edwin Southwood Jones was the manager at Cyrus Hanson's Henllis works in 1881, before establishing his own business in 1885. The Southwood Jones company produced firebricks until 1960, before being finally squeezed out by larger firms, particularly the Scottish firebrick conglomerates. (Lawrence Skuse)

The Stafford Coal & Iron Co. Ltd operated extensive collieries just south of Stoke-on-Trent. Stafford was one of the Great Fenton collieries, which the company established in 1873 to obtain the blackstone, ironstone and coal in the upper seams. The brickworks were situated north-east of the colliery towards Great Fenton and produced some very hard bricks. This example is a blue windowsill brick.

The Star Brick & Tile Co. Ltd operated several brickworks, including one on Llantarnam Road, Cwmbran, which is where this brick was made. Production here ran from before 1881 until at least 1958, and the company was still in business in 1973. In the 1920s red bricks and tiles were made from marls from the company's clay pity, while firebricks were made using fireclay brought by rail from the Varteg Deep Black Vein Collieries Ltd at Varteg, Abersychan. (Lawrence Skuse)

Sometimes things are not what they seem. The Stourbridge Fireclay Co. was so named because the fireclay discovered at the Porthmawr Adit, Upper Cwmbran, was of similar quality to that found in Stourbridge. Brickmaking commenced here in 1839 and later passed to R. Blewitt, who possibly operated under the name 'Cwmbran Fireclay Co.', Henry Parfitt, and finally Guest Keen & Nettlefold. The works closed about 1915. (Lawrence Skuse)

The Sunset Brick & Tile Co., Cowgate, Newcastle upon Tyne, was founded in 1921 by W. Cochrane-Carr. Shale was extracted from a nearby quarry using explosives until 1953, when a mechanical excavator and drag line were introduced. By 1967 the quarry was exhausted of shale and, with no nearby economic supply, the works closed. (Frank Lawson)

Smethurst Hall Brickworks, Rough Hill Lane, Jericho, Bury, Lancashire. Brickmaking started here around 1870, and in 1920 Richard Berry Ltd was raising fireclay from a drift at the bottom of the clay pit adjacent to the works. It later passed to Taylor's Bricks (Bolton) Ltd and the business was wound up in 1966.

The brickworks at Gerrans, north-east of St Mawes in Cornwall, was started by James Thomas around 1888. It was possibly established initially to provide bricks for extending his farm house at Methers Collyn, and was then continued for a few years under the management of a man named Alderman and his sons.

Joseph Timmis, Bradwell Wood Tileries, Chatterley, near Tunstall, Stoke-on-Trent. Timmis was already in business at Bradwell Wood in 1851, and the business continued under his name until 1889, when the proprietor was William Herbert Timmis. By 1900 it was Joseph Timmis & Sons and the business continued as a limited company until after 1924.

Situated on an inlet off the Carrick Roads to the east of Falmouth, these works ran from 1891 to 1907. The bricks were made using river mud and burned in two round kilns. In 1897 the works was owned by the Trelonk Brick & China Clay Co. Ltd.

Thomas Tunnadine was making bricks in 1871 at the Waterloo Brick Works, Pillgwenlly, Newport. Thomas Tunnadine appears by 1884, making bricks at Pillgwenlly, while Thomas was listed at the Malpas Road Works. By 1891 Henry was operating at Malpas Road, Newport, and Henry is not mentioned. He died in 1907 and, in 1920, S. H. and A. Tunnadine are listed at the Malpas Road works, which were taken over before 1926 by the Star Brick & Tile Co. (Lawrence Skuse)

J. & M. Tymm, Klondike Brickworks, Rose Hill, Marple, Cheshire. The works was opened in the 1880s to manufacture blue, red, and fire bricks, along with chimney pots. Production lasted until 1913, when the site was sold to Marple Urban District Council.

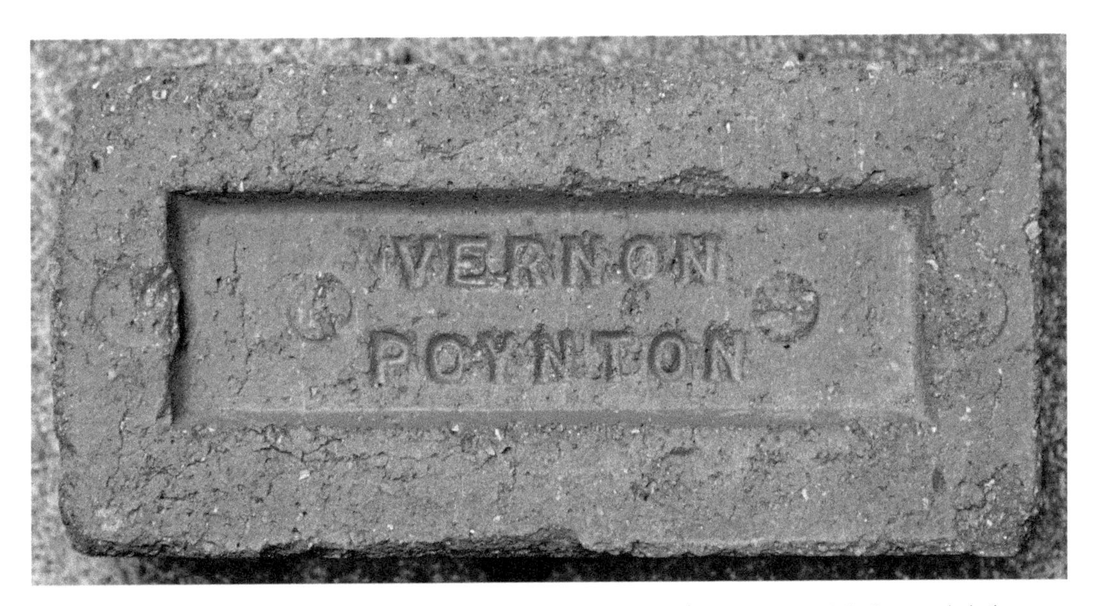

After the closure of the collieries of Poynton, Cheshire, in 1935, the Poynton Brick Co. used shale from the colliery tip to make bricks, using an open-topped kiln of the Scotch type. This was soon replaced by a large continuous Staffordshire kiln. The machinery was situated in the old colliery power house. By 1939 the works were run by the Vernon Brick Co., and soon after were taken over by J. & A. Jackson. The plant was electrified in 1956 and the kiln had a top put over it, but production ceased in 1958. Maximum output was 65,000 bricks per week.

Walsall Wood Colliery was sunk in 1874 on a site close to the Daw End Canal, just north of Coppice Road. A brickworks was established just to the north of the pit and a canal basin was constructed for brick traffic. There was a very large circular kiln with central chimney. The Earl of Bradford as mineral owner was paid a royalty of 1*s* per 1,000 bricks manufactured. The works are shown on the OS maps from just after the First World War, but the site had been cleared by 1938.

An unusual name for a brick manufacturer. This one is from the Ashgate area on the west side of Chesterfield. The works were established in the early years of the twentieth century and were closed by the 1950s.

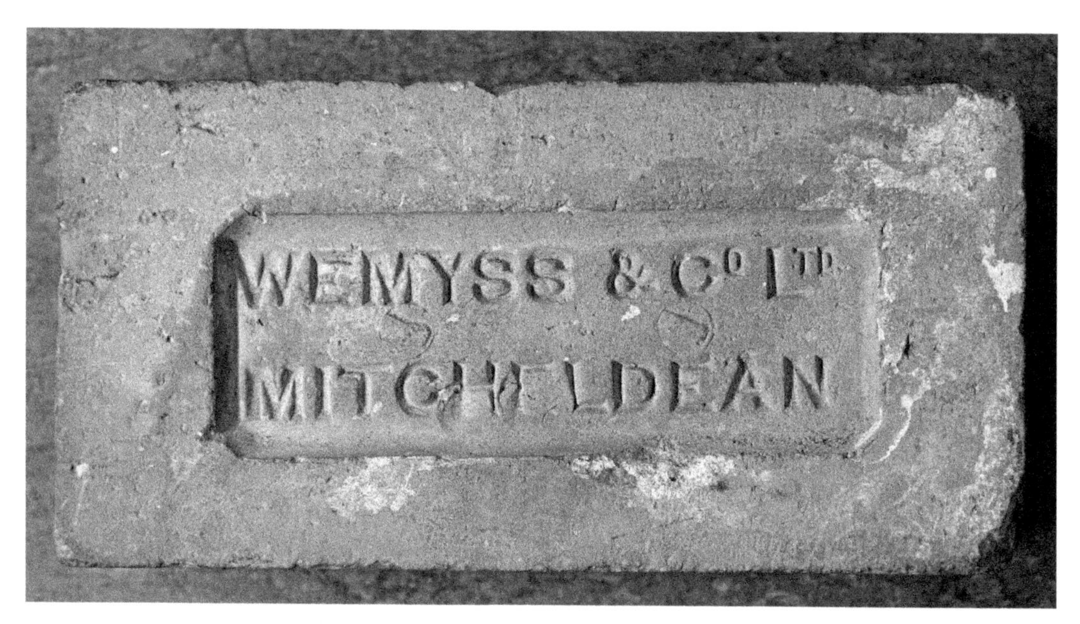

The Mitcheldean Stone & Brick Works (M. W. Colchester Wemyss with J. Miller Carr as manager) was opened in 1882. At its peak, the works employed some seventy people who, besides producing building stone and bricks, made urns, bowls, tiles, drainpipes, flower pots, garden vases, rustic stumps, pitchers, and architectural and fine art terracotta. There were three kilns, each capable of holding 65,000 bricks. In 1900 the Wilderness Brick & Stone Co. Ltd (Mitcheldean) was acquired by the Forest of Dean Stone Firms Ltd. The works closed around 1907.

This Staffordshire brickworks were adjacent to the West Cannock Colliery Company's No. 1 colliery, which commenced operations in 1869. Bricks were produced only for the company's use and were never sold outside. The plant comprised a Wooton Brothers brickmaking machine, capable of producing 28,000 common bricks per week and powered by a single cylinder steam engine. Two square kilns had 28,000 and 25,000 bricks capacity. In 1928 the plant produced 422,993 bricks but, in 1930, it was only 43,000 bricks and the works then closed.

Llandowlais Brick Works were run by Whitehead, Hill & Co., who took over from J. C. Hill & Co. in 1925. The works incorporated two eighteen-chamber kilns, each chamber having a capacity of 19,000 bricks. Output was around 304,000 bricks per week. Production ended after 1973. (Lawrence Skuse)

G. T. Whitfield opened the Whitfield brick and tile works on Robins Wood Hill, Tuffley, Gloucester, in the early 1890s. It was closed in the 1940s due to competition from larger brickworks.

This works was at Bridgrule, Holsworthy, on the border of Devon and Cornwall. Opening by 1898, it was sold in 1910 to the Western Counties Brick Co., who operated it as the The Whitstone Brick & Tile Co. Ltd. In the mid-1950s, production was around 4 million bricks a year using a round kiln and a nine-chamber tunnel kiln. It closed in 1965.

Woodside Brick Co. Ltd, Chesterfield Road, Woodseats, Sheffield. This was an extensive works and clay pit with three continuous kilns, which was operating by 1905. Nearly a million bricks were supplied from here to build the Sheffield Central Library and Graves Gallery, which opened in 1934.

Clement Wooldridge is listed in the 1867 and 1869/70 trade directories as operating at Davis Street, Hanley. The brickworks were probably a short-lived venture, and had a brick shed and two round kilns. Clement Wooldridge was born in Hanley in 1834 and is listed in the 1861 census as a joiner. Clement died in 1871 and the business was briefly run by his brother Levi Wooldridge, although his main occupation was as a victualler and pub keeper.

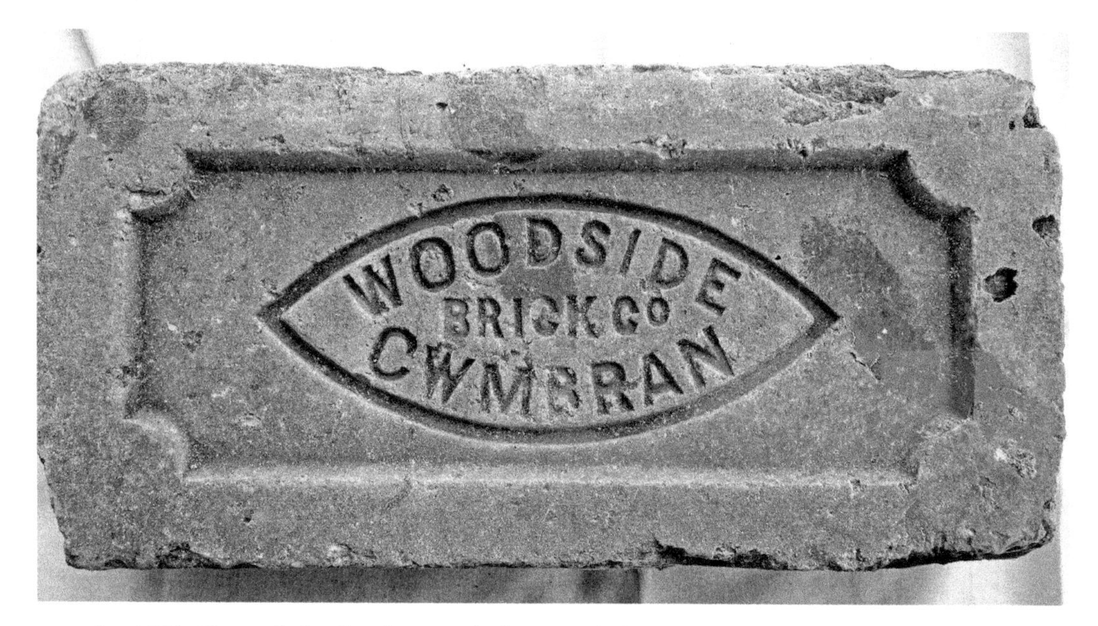

In 1875 Henry Bolt Sketch owned the Woodside Brick Co., Cwmbran, adjacent to the Monmouthshire canal. The Woodside works had crossed the canal by 1901, when the original site was derelict. I. W. Scourse & Son were proprietors in 1906 but, by 1914, the brickworks were listed under the Standard Brick Co. They had closed by 1926. (Lawrence Skuse)

The Ackton Hall Colliery Co. Ltd opened their brickworks near Featherstone around 1900. After 1947 control passed to the National Coal Board and then to The Midland Brick Co. Ltd (with several works in Nottinghamshire, Leicestershire and Staffordshire), which was absorbed by Butterley Building Materials Ltd in 1973. Closure of the brickworks seems to have occurred in the 1970s and of the colliery in 1984.

National Coal Board Bricks

One of the largest brick manufacturers in Britain was created by the nationalisation of the coal industry in Britain, which meant that many of the brickworks associated with collieries came into public ownership. At vesting day in 1947, the National Coal Board thus became owner of eighty-three brickworks and two pipe works, employing 2,943 people and producing about 10 per cent of the total British output of bricks. The brickworks varied widely in age, type and size. The largest, Niddrie brickworks in the Scottish division, produced over 25 million bricks in 1952, while twenty-nine works had outputs of less than 5 million bricks. Of these, three were in Wales, twenty in Scotland, and the rest in England. By 1951 the NCB was operating seventy-five brickworks, with a total output of 473 million bricks or 8 per cent of the national output in that year. New works were opened at Skares in 1954, Desford and Throckley both in 1963, and Callander in 1968, while some of the others were either modernised or closed entirely.

For the first fifteen years, each brickworks was the responsibility of the relevant colliery manager until it was realised that they were not necessarily qualified or interested in producing bricks. It was in 1962 that the National Coal Board Brickworks Executive was set up as a separately managed division. The Brickworks Executive lasted until 1971, by which time the remaining brickworks had either been closed or transferred to private ownership. Many of the Scottish works seem to have gone to the Scottish Brick Corporation in 1969.

Annesley Colliery was sunk in 1860 by the Annesley Colliery Company, with the first coal production in 1865. In 1925 the pit was sold to the New Hucknall Colliery Company. A brickworks with a twenty-chamber, modified Hoffmann kiln was established at the colliery by 1933 and continued under the National Coal Board until 1969. (Frank Lawson)

Birch Coppice Colliery and Brickworks, east of Tamworth, previously owned by Morris & Shaw Ltd. The colliery was sunk in 1875 and the brickworks first appear between the mid-1880s and 1900. It was expanded after 1947 by the National Coal Board but had been closed by the 1970s.

The Cramlington Coal Company built a brickworks with a large continuous kiln at East Cramlington Colliery in 1900. The company became part of Hartley Main Collieries in 1929 and the pit reopened after being closed since 1922. Cramlington brickworks became part of the National Coal Board brick business and, in 1963, was producing pressed commons and pressed rustics. (Frank Lawson)

The Waterside brickworks of the Dalmellington Iron Company at Dunaskin opened in 1928. In 1931 it was taken over by Baird & Dalmellington Ltd. This works featured a Staffordshire transverse arch kiln that had fourteen connected chambers, which were heated to 1,100–1,200° C by an underfloor heating system. The draught from the chimney drew heat through each chamber and up to 16,000 bricks were fired for two weeks in each chamber. In 1947, ownership passed to the National Coal Board and later the Scottish Brick Corporation until closure in 1976.

Fauldhead brickworks produced pressed commons, burned in a Hoffmann continuous kiln. The National Coal Board operated these works from 1947 until shortly after the closure of the colliery in 1969.

Hednesford brickworks were originally owned by the Cannock Chase Colliery Company. It was taken over from the NCB in 1974 by the Butterley Brick Company and closed some years later. It produced a wide range of bricks, including plain red pressed facings, and engineering bricks in brindle, brown, and blue.

Coppull Colliery was started by Thomas Rymer Bourne in the 1860s. It was renamed Hic Bibi Colliery by 1871 and had several owners until it closed in the late 1880s. The Ellerbeck Collieries Company continued to lease the adjacent large brickworks, which eventually came to the National Coal Board, who ran the business until closure in 1958.

Hickleton Main Coal Company sank their colliery at Thurnscoe, near Barnsley, between 1892 and 1894. By 1906 there was a brickworks to the south of the pit, but this had gone by 1930 and was replaced by a larger works to the north with a single Hoffmann kiln. Ownership passed in 1937 to Doncaster Amalgamated Collieries Ltd, and then the National Coal Board in 1947. In the last years up until the closure of the works in the 1960s, the bricks were just stamped 'HICKLETON'. (Frank Lawson)

Hilton Main brickworks comprised four continuous kilns, situated just south-west of the shafts at Hilton Main Colliery. The National Coal Board continued to produce bricks here after nationalisation until 1967. The colliery closed two years later.

Kirkby Summit Colliery was developed by the Butterley Company from 1887, although initially it did not have its own brickworks. However, large works with two continuous kilns appear on the site during the 1930s. These works produced pressed commons and lasted until the closure of the colliery in 1968. (Frank Lawson)

The Mickley Coal Company operated a colliery at Stocksfield, 8.5 miles east of Hexham in Northumberland. The National Coal Board continued to run the brickworks, although the colliery was closed about 1947.

Mitford was a trade name for the bricks produced at Blaydon Brickworks, Blaydon-on-Tyne, County Durham. Blaydon firebrick works were one of two originally operated, along with Blaydon Burn Colliery by Joseph Cowen and his sons from the early nineteenth century. The National Coal Board operated the lower works, which had a system of railway sidings and a long stretch of wharfage for transport by rail or boat. (Frank Lawson)

Astley and Tyldesley Collieries worked Gin Pit, Nook Pit and St George's Colliery, and the brickworks was situated at Nook Colliery. Production ceased before the formation of the Brickworks Executive in 1963.

By 1874, bricks and tiles were already being made at Prestongrange Colliery. Downdraught kilns for burning bricks and pipes survived until a Hoffmann continuous kiln was built in 1910. This was replaced in 1937 by another Hoffmann kiln, which remains today. The colliery continued to supply raw materials for the brickworks until its closure in the early 1960s. A steady decline of the brickworks thereafter led to it closing in 1975.

The brickworks at Roslin Colliery were established by the Shotts Iron Company in 1937. It continued in NCB ownership until 1969 and then under the Scottish Brick Corporation until 1977.

Sherwood Colliery in Mansfield Woodhouse was sunk in 1902. Brickworks were working in the pit yard by 1914 and the continuous kiln remained in use into National Coal Board ownership. Closure of the brickworks was on 19 April 1958, although the pit remained open until 1992.

The Upton Colliery Co. Ltd was formed in 1923; the first coal was raised from the colliery in 1927 and brickworks with a large Hoffmann kiln were at work by 1928. After passing into National Coal Board ownership, the brickworks were closed by 1963, while the colliery followed in 1964. (Frank Lawson)

Watnall colliery was sunk in 1873 in Eastwood, Nottinghamshire and, while brickworks were operating on the site by 1900, the main works were not started until the 1920s by Barber, Walker & Co. Ltd. They used the colliery shale, ground to powder, moistened, and pressed in Bradley & Craven stiff-plastic presses. The colliery closed in 1950 but the brickworks continued at work. The kilns comprised a modified Hoffmann kiln, a Manchester kiln, which held 12,000 bricks per chamber, and, later, two Staffordshire kilns, which were built by the National Coal Board and held 20,000 bricks per chamber. Production ended in 1969.